Nature Style

Nature Style

Cultivating wellbeing at home with plants

—

Alana Langan & Jacqui Vidal
of Ivy Muse

Photography by Annette O'Brien

CONTENTS

01.	Introduction	7
02.	What is biophilia?	13

- Biophilic design
- Benefits of biophilic design

03.	Biophilic styling principles	25

- Houseplants
- Views
- Materials and decor
- Colour
- Shapes and forms
- Ambience

04.	Styling solutions for every room	79

- Kitchen
- Living room
- Bathroom
- Bedroom
- Retreat space

05.	Further reading	149
06.	Plant care	153

01

Introduction

IN SPRING 2014, we launched Ivy Muse, a botanical wares design studio based in Melbourne, Australia. What started as an idea between two long-time friends to encourage creativity with greenery has grown into an all-encompassing commitment to enhance wellbeing through the use of plants. Alana's background as an interior stylist working with some of Australia's best-known magazines and brands and Jacqui's eye for aesthetics as a former gallerist informs our shared passion for design and our approach to plant styling.

When we began, houseplants were a fringe accessory, not the 'decor du jour' they are now. Only the bold would entertain the idea of keeping plants indoors. However, in recent years we have seen this movement evolve and gain momentum, as people become increasingly aware of the health and wellbeing benefits plants can bring. We couldn't be happier that keeping houseplants is no longer a trend but a way of life.

We've been fortunate to work with many people and businesses to plant style their homes or offices over the years. Plants have become a bona-fide element in the interiors world and we love to share our expertise in this field, often collaborating with designers and architects on projects. It is through this work that strengthening the connection to nature in homes became an organic extension for us.

With so much time spent indoors, it is vital that our homes nurture us and support our wellbeing. Amid today's fast-paced society with the daily barrage of blue-light technology, the need to tap into nature's restorative powers is greater than ever. The more high-tech our lives become, the more we need nature. Nature can help humans to thrive, promote health and wellbeing, and increase creativity and mental acuity. It alleviates our stresses, reinvigorates our senses, rejuvenates our bodies and minds, and restores our energy. But if you don't have the luxury

Alana Langan (left) and Jacqui Vidal (right)

of the forest at your doorstep, incorporating the elements of nature into your home is definitely an accessible alternative.

This book is about taking the theory of biophilic design and distilling it into easy-to-follow steps that can be applied in the home, both economically and effectively. We call it 'biophilic styling'. You can use this book to help you to make decisions on indoor plants and how to style with them, or to guide choices about materials, decor, finishes and furnishings. If you want to incorporate more of the natural world into your home, pick and choose the biophilic elements that work best for you. You don't have to be renovating or building from scratch, and it makes no difference if you live in a small city apartment or a rambling home in the country.

We have included both our homes in this book. They are vastly different in style and location, illustrating the breadth of biophilic design and styling and how it can be implemented in a multitude of ways. We hope you are inspired, just as we have been, to live in synchronicity with the natural world and to restore yourself while doing so.

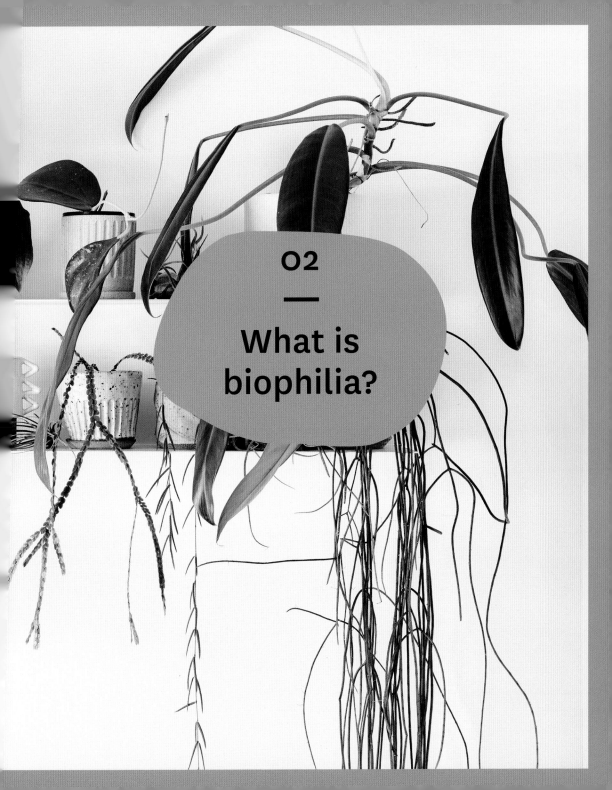

WHEN YOU THINK OF NATURE, how do you feel? Relaxed? Invigorated? Happy? Most people don't need convincing to get out into nature. Without even realising it, we're inherently drawn to the natural world. That's biophilia.

The word 'biophilia' was first introduced in the 1960s by social psychologist Erich Fromm and later popularised by biologist Edward Osborne Wilson in his 1984 book, *Biophilia*. Wilson explored the theory that humans have an innate desire to connect with other forms of life; a deep-seated affinity with the natural world that has built up over the course of human evolution. Biophilia can be summed up as an intuitive love of life and living systems.

Biophilia encourages us to slow down and pay attention to our surroundings. Maybe you can smell wisteria on the breeze through an open window, or feel the warmth of the sun on your skin. Perhaps you can hear a chorus of birdsong outside, or notice the calming effects of a walk in the forest. In its multitude of forms, biophilia soothes us deep within.

BIOPHILIC DESIGN

Biophilic design is a practical way people can connect with nature within the modern, built environment. It recognises that natural systems and processes, such as the passage of daylight or good ventilation in buildings, are critical to human performance and wellbeing. Stephen Kellert, an American professor of social ecology, pioneered the contemporary concept of biophilic design. In the 1980s, he and Wilson began collaborating on a series of articles and books, which became the foundation for Kellert's innovative work.

However, it's only in recent years that biophilic design as a concept has become more popular, bolstered by the global trend towards wellbeing and eco-conscious living, although biophilic design has been with us for centuries. From the architecture of ancient Greece, which incorporated natural design patterns, to the Neolithic villages of Scotland with their houses and furniture crafted from stone, integration with the natural environment, as well as use of local materials, has long been a favoured practice.

Combine a contrasting mix of plants like *Hoya heuschkeliana*, chain cactus and variegated inch plant to give your shelves a wow factor with interesting form and depth.

Biophilic design has three overarching elements: 'direct' connection to nature, 'indirect' connection to nature and 'experience of space and place'.

'Direct' refers to direct experience. Think of the feel of the sun on your skin, smelling the fresh spring breeze through your open window, caring for your houseplants or warming yourself by a crackling fire.

'Indirect' connection can be anything that evokes a sense of the natural world, particularly natural materials, colours, forms and shapes. Artworks or photos of nature are also considered indirect, as are textures and patterns that may be found outside. Even elements and finishes that age and show a patina – for example, raw brass and copper – can be considered biophilic, as they represent the passing of time, a natural occurrence.

'Experience of space and place' refers to spatial features that are reminiscent of those found in nature. Examples of such features include outdoor views, visual connections between interior spaces, and cosy spots that offer refuge and attachment to the environment. This element is generally incorporated into the initial design of a space, but there are still some key tactics you can try out at home, which is where houseplants really come into play.

BENEFITS OF BIOPHILIC DESIGN

Our mental and physical wellbeing is closely linked to our relationship with the natural environment, so if we want to achieve healthy, satisfying lives, interacting with nature is a necessity rather than a luxury. Biophilic design helps us to connect to the natural world by engaging our senses in a way not offered by technology and the built environment.

A raft of growing scientific evidence from around the globe supports the notion that biophilic design is beneficial to human health. Leading global trailblazers like Apple, Google and Amazon have invested heavily in biophilic design in their HQs. Office design that incorporates biophilic principles is now recognised as a key factor in improving staff engagement and productivity, reducing absenteeism and increasing overall worker retention.

Incorporating biophilic attributes in your home can bring a multitude of calming and restorative benefits. From lowering blood pressure, stress and irritability to boosting self-esteem, enhancing mood and encouraging healthy lifestyle habits, connecting to nature offers a plethora of wide-ranging physical and mental advantages. In fact, a landmark study in 1984 found that contact with nature enhanced the healing process: hospital patients with views of nature recovered more quickly when compared to those patients with no views of nature.

Houseplants are an essential part of biophilic design and will provide a multilayered experience of nature – you can touch them, see them, even smell or taste them. As well as being one of the quickest and easiest ways to connect with nature indoors, houseplants offer other benefits. Tending to them will encourage you to slow down and be present, and you will find that the practice of simply caring for another living thing is deeply rewarding in itself.

Biophilic design is most effective when attributes are layered, creating multisensory encounters that engage sight, hearing, touch, smell and taste. The goal is to actively increase your direct and indirect connection with nature, and to enhance how you experience your home.

Open the doors or a window and listen to the birds. Maybe you can see the breeze blowing your curtains or hear leaves rustling nearby. Perhaps you can smell freshly cut grass from outside or flowers blooming by the window. Natural ventilation from open windows or doors helps productivity, with even the smallest amount of fresh air proving beneficial. Growing herbs in a sunny spot indoors, tending to them often, enjoying their aroma and cooking with them is a year-round activity. Surrounding yourself at home with a plethora of plant-based elements can be a hugely rewarding endeavour, and from this enriched place you can move on to layering your connection to nature.

Create a cosy nook to retreat to in any room with your favourite comfy chair and a striking, large plant like an umbrella tree. Spots like these can help you feel secure and cocooned when you want to take refuge and have a little time out.

Open windows provide a connection with the outdoors by engaging multiple senses at once. You may be able to smell the delicate scent of flowers in bloom nearby or feel a cool breeze on your skin.

03
Biophilic styling principles

BIOPHILIC DESIGN IS WIDELY RECOGNISED by architects and interior designers, but you don't necessarily need to be building or renovating to introduce the concept into your home. Biophilic styling is affordable and accessible, and can be incorporated into most spaces. The key is a holistic, layered approach that will engage the senses profoundly. It's about making the most of your unique space by paying attention to the surrounding natural environment, the period of the home and how you and your family use it. Whether you keep things simple with a few potted plants here and there, a wall painted green or a nature-based artwork or two, or create the ultimate cocoon of natural elements and plant life, trust your instincts and go with what feels right.

We have distilled our approach to biophilia into six key styling principles that can easily be incorporated into an existing space:

- HOUSEPLANTS
- VIEWS
- MATERIALS AND DECOR
- COLOUR
- SHAPES AND FORMS
- AMBIENCE

If you're in the process of designing or renovating your own home, you can also utilise these suggestions and make them part of your planning from the outset.

A textural rammed-earth wall makes a great backdrop for a lush indoor vertical garden dripping with an array of plants, including hoya, donkey's tail and devil's ivy.

HOUSEPLANTS

If you don't have natural vistas or garden views to enjoy, create your own jungle inside. Houseplants are a powerful element in your toolkit and are often the easiest and most cost-effective way to make a space biophilic, especially when you don't want to change your interior decoration or furniture. Plants soften interiors and make them more inviting, transforming sterile spaces into sanctuaries. They really are the ultimate indoor accessory.

Create a focal point by placing a large potted specimen like the never never plant on a table.

Houseplants offer a huge number of benefits. They are excellent humidifiers and refresh the air by drawing in carbon dioxide and releasing oxygen. Just as nature does, plants can help advance your health and wellbeing. Be warned though, they're addictive!

Having just one or two plants won't have much effect. Yes, you heard right – plants should be abundant indoors! As well as choosing a variety of plants that speak to your taste, where possible incorporate plants that are native to your area. Foliage or flowers from your garden or suburb – or even the local florist – are another great way to add a touch of nature inside, although they don't last as long as potted plants.

Incorporating plants into the home may feel like an overwhelming endeavour, but once you get to grips with the basics, it's really easy. Anyone can transform their home into a lush, green oasis. Remember, when it comes to plants, the more the merrier. Don't be afraid to layer up your specimens, and choose plants that you love rather than sticking to any strict style. But remember: plants need a suitable environment to thrive, so always pay attention to each plant's exact needs when assessing its suitability for your home.

VIEWS

Sometimes the simplest way to connect with nature is to look outside. You will see layers of earthiness – a literal multitude of elements for your senses to experience. For this reason, keeping clear, unobstructed views to the outdoors is key. This is a wallet-friendly and very simple biophilic hack that shouldn't be overlooked.

Houseplants are a key element of biophilic design. Combining different specimens, such as fiddle leaf fig, long leaf fig and hoya, will help replicate the variety of plants found outdoors and create seamless views of nature from inside to out.

If you have views to appreciate, pull back curtains or open blinds and shutters where possible, and let nature and natural light reign supreme. Do you have any solid doors that could be changed to glass? Or frosted glass that could be replaced with clear? Skylights can be a great way to create sightlines to the outside and houseplants love sitting under them too! Also consider your window treatments. Can you replace heavy curtains or blinds with sheer fabrics? Or even remove window coverings entirely?

When you've been living in the same space or with the same furniture for a while, sometimes you don't realise how it can be improved. Take time to step back and review the layout. Are all sightlines to the windows and doors clear? Do you have a wide or tall plant that's blocking a window? Try to find a more suitable spot where it can thrive, although not at the expense of your outlook. Position sofas, chairs and other seating to make the most of the view, and ensure there are comfy nooks or places to which you can retreat and enjoy your surroundings.

Place a comfy chair by a window and enjoy the ambience. Create lush scenes inside with large and wide-reaching plants such as the kentia palm and philodendron 'Hope'.

Position plant life throughout a room to create a flowing scene you can enjoy from any vantage point. Hurricane cactus teamed with a brass pot creates a striking sculptural form, while heartleaf philodendron and blue star fern complete the setting.

Using mirrors can enhance your outlook as well. Mirrors provide a sense of space and reflect light, helping to create a bright, warm environment. Whether it is a small wall-hung mirror or a full-length, floor-standing one, use its reflective surface to capture a picture of the outdoors. And here's a top tip: position your favourite plant in front of a mirror. It will reflect the light onto the rear of the plant, which will help its foliage grow evenly.

Earthy, peach-toned walls set the scene for a plant-based vista at the end of this hallway. Create layered green vignettes using a mix of plants of different heights or combine with plant stands. Such variety in your display will echo scenes reminiscent of nature.

If you're lucky enough to have flexible living and dining spaces, position furniture so that you can enjoy views as much as possible. Instead of using cumbersome furniture and lighting, select slimline options that won't block vistas. Perhaps you have a butcher's block that can face a window so you can enjoy the outlook while prepping dinner? Or maybe you have an unused corner of your kitchen that could be transformed into a small breakfast bar complete with a view?

If you cannot gaze upon nature from inside your home, having indoor plants is the next best thing. They can play a huge role in making a space more biophilic. You can actually create views inside the home with houseplants. Group plants together to create a focal point within a space. Position them where they can be appreciated from a distance – for example, at the end of a hallway – or turn boring corners into a plant-gang haven. Place plants throughout your home so that each room has a consistent theme of greenery. Even if you do have views of nature outside, this will help connect to those views even more. And if you don't, you still get the benefit of experiencing all that plant life!

MATERIALS AND DECOR

Natural materials and decor that show signs of age, changing and developing a patina over time, create a fundamental link with the outside world. Artificial products rarely evoke that sense of deep-seated satisfaction we get when we see organic forms and materials with a tale to tell. Think about how you feel when you come across a well-weathered leather chair, complete with the markings of a life well spent, or a brass bowl that has been a home for trinkets or keys for years. The pleasure and sense of connection we experience when we see and handle such objects is biophilia.

Our senses thrive on variation and contrast. Choose a mix of materials and finishes that will engage your senses, from smooth timber grain to textured ceramics and rough seagrass to polished copper. The organic, comforting and familiar appeal of natural materials soothes both mind and body. The following list is by no means exhaustive but includes some of our go-to materials that will give you a good starting point.

HANDMADE CERAMICS

There's nothing quite like handmade ceramics, with their undulating curves, irregularities and one-of-a-kind charm. The ethos of the maker is often woven subtly into each piece, telling its own unique story. These days, ceramics are not only reserved for tiles, tableware, pots and trinkets, but also used for lighting, art and more.

Plants and handmade ceramic pots were made for one another. The fun of pairing your favourite green friend with its potted home is almost as enjoyable as choosing the plant itself. The organic nature of such pots, and the multitude of finishes, colours and shapes, offers the perfect accent to all manner of plant life, with each piece as distinctive as the specimen that calls it home.

WOOD

Often regarded as one of nature's most adaptable materials, wood offers a veritable smorgasbord of uses, with its diverse types, finishes and qualities adding character and warmth to spaces. From its use in wall treatments, flooring, joinery and furniture to smaller decorative touches such as bowls and utensils, wood brings a comforting ambience to a space. Not only practical, it also ages gracefully.

Incorporate hand-carved timber stools as plant stands. Invest in wooden planters, which come in a variety of sizes from small tabletop options to those suitable for large trees. Small fallen tree branches or twigs make great sculptural elements and can also be used as an organic base to display air plants.

BRASS AND COPPER

Brass and copper are two of our favourite biophilic elements. As they develop a patina, each with its own unique markings, they reveal the passage of time, which is something we, as humans, seem to innately connect with. These metals, often made from scrap elements, can be used in a multitude of ways, from joinery details to smaller items like tableware, cutlery and even handles.

    Brass and copper planters complement green-leafed friends well, and the contrast of soft foliage with the lustrous shine of the pot can be eye-catching. Don't be restrictive and stick to one type of metal pot either. We love to mix brass and copper planters throughout a home for a refined take on a little indoor-plant bling.

WOOL

Cool in summer and warm in winter, wool showcases nature's adaptability at its best. Wool carpets and rugs, known for their luxurious feel and stunning visual appeal, are organic in both body and texture. Their sustainability makes them an ideal choice when considering floor coverings for living and bedroom areas. For a more affordable option, sheepskin rugs come in all manner of sizes and are a great addition for instant comfort and warmth. Not only for floors, they can also be used on sofas, chairs and beds. The soft, thick texture of wool contrasts well with the cool foliage of houseplants and can be a lovely addition to any biophilic dwelling.

STONE

Cool to the touch yet grounding and elemental, natural stone refers to a number of products quarried from the earth, such as marble, granite, travertine and slate. The beautiful imperfection of natural stone is undeniable and always a pleasure to look at and feel.

Carved marble planters are a unique option for houseplants and work best for small- to medium-sized specimens. Any larger and they can become very heavy and cumbersome. From the more subtle white and grey tones to daring, bold shades like pink and green, there's a colour and variety of stone to suit every room.

LINEN

A natural and biodegradable textile spun from fibres of the flax plant, linen offers a luxurious yet down-to-earth feel. Its texture and tactility engage our sense of touch and can have a calming effect, especially helpful at bedtime. It is a breathable fabric so it keeps you cool in summer and warm in winter, and it's also hypoallergenic, naturally repelling dirt and bacteria. Like all good things, it gets better with age; each wash softens it.

SISAL, JUTE AND COIR

Natural fibres such as sisal, jute and coir are spun from plants. Hard-wearing and functional, they are great when used for flooring and rugs, and kitchen accessories such as placemats and trivets. Coir poles can also be very effective when managing unwieldy plants like devil's ivy, fruit salad plant or philodendron.

COLOUR

Colour can have a profound effect on how we feel in a space. It is also one of the most significant ways to make a dramatic shift in an environment. The way we experience colour is hugely subjective. For one person, blue may remind them of the ocean and days spent with family by the beach on summer holidays. Another person may be reminded of a place or memory they'd rather forget. So not only does subjectivity come into play, but the way we interpret colour through our sight and cognitive functions also has an impact. This means that successful use of colour comes down to one thing – doing what feels right for you.

From a biophilic perspective, the introduction of earthy colours that represent nature, especially shades of green, is thought to have a grounding, comforting effect. However, if green isn't your jam, and you think brown should really just stay in the 70s, then work with shades that are still reflective of Mother Nature but a little more in line with your taste. Try terracotta, pale pinks, stone shades and neutrals.

Team eye-catching houseplants such as philodendrons, chain cacti and blue star ferns with textural ceramics to create an earthy colour palette with a soothing effect.

Intuitive application of colour in your home may not necessarily win you any awards or popularity contests, but by using it wisely, you can hope to enhance your comfort and happiness and ultimately your wellbeing. A fresh coat of paint is one of the cheapest and most effective ways to introduce colour. Think beyond walls and ceilings and consider painting furniture, cupboards, or even tiles and floors, to make your space more to your liking. Creativity reigns supreme here.

Don't be afraid to combine black with complementary shades of green when it comes to pots and plant stands. Using colour in this way will draw attention to your hero plants, such as long leaf fig and fruit salad plant.

Another great way to introduce colour is with houseplants, especially if you're keen on a monochrome or minimal approach when it comes to indoor decoration. Plants come in a huge variety of nuanced shades of green, so one option is to layer the depth and variety of foliage, just as you might find in nature. You can also take a bolder approach and choose plants that have more dramatic colours, like rubber plant 'Tineke', which includes cream, green and dusty pink on its broad leaves, or purple shamrock, with its eye-catching deep-burgundy foliage.

Don't forget that the colour of your planters can also create an effect. This is another great opportunity to enhance your biophilic experience. Planters in shades reminiscent of nature – and especially those with organic handmade finishes, materials and glazes – make an ideal pairing with plants. If you're not feeling adventurous then play it safe and stick with green. You've already introduced the colour with your plants anyway, so adding different shades with your planters won't seem jarring or out of place – they'll simply blend in with the tonal shades of plant life you have already established. Another safe choice is to use neutrals such as black, grey or white.

SHAPES AND FORMS

It only takes a quick glance outside to see that curves are abundant in nature. Straight lines are often the work of machines and the modern world. In contrast, the human form exemplifies curves, and the furniture or decor in our homes often reflect this. Whether it's the smooth edge of a round table or a handcrafted bowl, softer lines welcome our touch.

Grape ivy is a fast-growing, low-maintenance plant with cascading stems that can be trained vertically. Full and luscious, this versatile all-rounder with glossy, dark-green foliage is great for 'green shelfies', tabletop styling or console displays.

Rabbit's foot fern, coin leaf peperomia and air plant

Many people feel drawn to curved lines, shapes and objects. Fill your home with a variety of pieces inspired by the natural world – think circles, curves and waves – or take inspiration from trees, leaves and even animals. When it comes to indoor plants and their accessories such as plant stands, plant stakes, pots and vessels, the same rules apply. Instead of going for angular, rigid designs, opt for more organic and shapely forms. You can also look to the plants themselves for inspiration and create pairings between planter and foliage that are complementary. For example, try planting a coin leaf peperomia (with its heart-shaped leaves) in a bowl-shaped pot, or pair a sculptural grape ivy with a rippled ceramic pot that echoes its unique form.

AMBIENCE

Often intangible, yet always profoundly felt, ambience refers to the atmosphere or 'feeling' of a space. A range of elements combine to create the mood of a room, most notably light, air and temperature, and sound and scent. Incorporating these natural elements in the home can create a comforting, nurturing environment that engages the senses and helps build a connection to the world outside.

LIGHT

Exposure to daylight is vital for wellbeing. It helps our circadian rhythm (a biological process that determines when we sleep and wake) and also enhances our productivity levels and mood – we feel better and more energised when we experience it.

Varied light throughout the day, and the shadows created within the home as the sun rises and sets, helps us connect to the cycles of nature. Where possible, keep curtains or blinds open. Not only is daylight good for you, it's a necessity for your houseplants.

Mistletoe cactus looks particularly impressive displayed in hanging planters, where its beautifully abstract tendrils can descend at length. It prefers a bright, sunny spot – next to a window indoors is ideal.

Long leaf fig, organic ceramics and candlelight from a brass oil burner create a gentle biophilic effect. Long leaf fig prefers a bright, sunny position but can tolerate lower light levels.

The majority of indoor plants grow well in bright, indirect light, although some prefer a sunnier spot, while others can tolerate shadier positions. The key is to identify the orientation of your windows. In Australia, north-facing windows get the most light during the year, while in the northern hemisphere, the opposite applies and a south-facing position gets the greatest light. Work out how light moves through your home and at what intensity (for example, afternoon sun is much stronger than early morning sunlight). You can then identify which plants are best suited to those positions.

Candlelight is a great way to set the mood at night and further support the natural rhythms of the outside world. Rather than having a ton of artificial lights on when the sun has set, candlelight offers a softer, less intense way to light your evening.

AIR AND TEMPERATURE

Subtle changes in airflow and temperature, like a gentle breeze passing across our skin or the warmth of the sun through a window, are important considerations in any biophilic space. Surfaces which vary in temperature – think cool marble or ceramic, warm leather or sheepskin, a hot cup of tea or an open fire – mimic natural environments and offer us comfort.

Plants like philodendron 'Hope', fruit salad plant, swiss cheese vine and rubber plant prefer a consistent room temperature.

Eye-catching tassel ferns require good airflow to thrive. Position them on shelves or plant stands where their long tendrils can hang down to catch passing soft breezes.

Houseplants are a great tool for creating optimal air quality and temperature within your home. Not only can they purify the air by filtering out toxins, they also add moisture and humidify dry indoor air, which many people find an irritant, especially during winter. Your plant gang can also be a useful barometer for air quality. If you're regularly finding dry, brown tips on their foliage, it may be a sign the air is too dry in your home – for them and for you!

Showcase impressive plant life like philodendrons, fruit salad plants and fiddle leaf figs indoors to create a striking plant gang. As well as helping to detoxify the air, they can create a seamless connection to the landscaping outside.

To increase variations in airflow and temperature, open doors and windows and connect with the outdoors. If it's a chilly day and opening the window isn't ideal, lighting a fire indoors may be an option. Too hot outside? Turn on a fan and feel the breeze on your skin. Too much time spent indoors, in artificial environments devoid of fresh air, can deplete our energy levels, negatively affecting concentration, productivity and an overall sense of contentment.

Fresh air goes hand in hand with good health. With so many toxins commonly found in built spaces, as well as the furniture and other objects we add, proper ventilation is essential to rebalance mind and body. When the air in your home is stagnant and chock-full of toxins, the air outside might be better for you. If you're living in a heavily polluted city, be sure to take regular trips to the forest, bushland or another green space to help recalibrate your body.

Variegated creeping fig and peace lily create a green vignette. Plants can help hide more unsightly and functional objects, such as diffusers or appliances.

SOUND

When immersed in nature, the sounds all around us are very different to those we hear inside, such as the mechanical ticking of a clock or the hum of a fridge or an air conditioner. In your home, prioritise the echoes of the outdoors over urban sounds to support a seamless connection to the outside world. This can often be as simple as opening a door or window, listening to the birds chirping, the leaves rustling in the breeze, rain falling or waves crashing. If the environment outside your home isn't therapeutic, focus your efforts on what you can achieve inside.

The sound of water gently flowing is a powerful biophilic element (although not always practical when inside the home). One easy alternative is to use an essential-oil diffuser. They often make a gentle, trickling water sound while they disperse a fragrant aroma into the room.

SCENT

Scent and our sense of smell are powerful tools. Every breath is a sensory opportunity to submerge ourselves in nature. Our sense of smell starts developing inside our mother's womb, indicating the influence scent will have on our consciousness throughout our lives. Intrinsically entwined with our emotions and memories, the power of scent can be harnessed to create a state of oneness with nature.

Modern homes can be awash with artificial aromas from cleaning products, household toxins and strong perfumes that pervade our environment. To get back to basics, use unscented natural products as much as possible and, from there, introduce fragrances from the outdoors in a more refined and subtle way, reflective of the mood you want to create.

Position aromatic flowers or plants like lavender, rosemary and basil in garden beds by windows, or add indoor herb planters. Fresh flowers picked from the garden, such as gardenias or roses, make a fragrant addition to any room and immediately increase your direct connection to nature. Scented room sprays infused with essential oils from the bush, such as lemon myrtle and eucalyptus, can be uplifting and luxurious.

With a little patience, and optimal conditions like good light and humidity, plants such as krimson princess hoya will flower when mature. They produce delicate clusters of pink, star-shaped flowers during spring and summer.

Peppercorn plant teams with fragrant palo santo and handmade incense to deepen the connection to nature, creating a comforting, layered scene that encourages you to slow down.

Houseplants like hoya, with their sweetly scented wax-like flowers, are a beautiful way to add fragrance to the home. However, they do need at least three to four hours of direct sunlight daily to flower, a process which can take years in some cases (but is always worth the wait). If you've got a very sunny spot, indoor citrus trees such as the Meyer lemon are also a great option. With a little luck and patience, you might even be able to enjoy homegrown fruits from them.

Scent can be used to advantage in any room – consider using centuries-old traditions like incense, wood or dried-herb burning to infuse your surroundings with the aroma of nature. Focus on scents that evoke memories of time spent in the outdoors. Try botanical notes of sage, wattleseed and juniper berry alongside more earthy foundations of cedarwood and rosewood.

04

Styling solutions for every room

Plants are a quick and easy way to increase the biophilic effect in a room. Use foraged foliage like rough tree fern, olive tree or even herbs like sage with complementary indoor plants such as swiss cheese vine and grape ivy.

FROM A DESIGN PERSPECTIVE, if you have the luxury of creating your home from scratch, you can use a wide range of natural materials and really go to town and bring your dream biophilic space to life. If that's not an option though, don't fret. There are still countless ways you can tweak your existing home to be more biophilia-friendly, no matter what the current look and feel of your space, its size or your budget. Time to have some fun!

KITCHEN

Kitchens are one of the hardest-working places within a home. They need to be functional yet also warm, welcoming and nurturing. Generally considered the heart of the home, the kitchen is where we often find ourselves, whether it is spending time with family, working from home, entertaining, or preparing and cooking food. Plants, both indoors and out, can help you achieve a biophilic space and are the easiest and most effective way to connect directly with nature in your kitchen. For this reason alone, it makes so much sense to incorporate them here – and as many as possible!

Not only can you literally surround yourself with houseplants and incorporate views to nature and organic elements in this space, but you can also really engage your senses in a number of beneficial ways, including eating your own homegrown produce. You can cook a plethora of plant-based meals using fruits and vegetables, create herbal tinctures or enjoy tea rituals to nourish and nurture yourself and others. You can grow herbs in the kitchen, appreciating not only their beautiful aromas but also their added flavour and health benefits by incorporating them into your cooking. You can even forage for plants or herbs in your local environment.

Now, we realise most kitchens have a general layout of benchtop and cabinetry, which means it can be hard to change things too much without actually renovating. But there are definitely ways to make your kitchen biophilia-friendly.

Consider placing a favourite pot-and-plant pairing in the kitchen. Long leaf fig teamed with a handmade ceramic pot is a winning combo.

HOUSEPLANTS

The kitchen is a great place for your leafy friends to call home. As a place of constant activity, it offers a special opportunity to check on your plants regularly – a key factor in helping them to thrive. Pay attention to what's happening to them. Are the leaves turning brown or yellow or looking limp? Are there any signs of pests? Is the soil dry or moist? When you keep abreast of how your plant is doing, you'll be much better equipped to recognise any signs of distress and be able to rectify the problems quickly before your plant is damaged beyond repair.

Including plants like peace lilies or spath sensations in the kitchen is a great idea, especially if you share a house with others. With their big broad leaves, these wondrous plants can help absorb the sound of footsteps and other activity when placed in pots or vessels on the floor. They're also great for cleaning the air and help to soften all the straight lines often found in kitchens. A win all round!

When styling with houseplants, think about the unique spaces your kitchen offers. What interesting features does it have? Do you have lots of shelves you can fill with plants? We love nothing more than creating 'green shelfies' with a mix of objects, ceramics and trailing plants. Maybe there are original details like ceiling beams where planters and lofty Boston ferns could hang. Is there a giant island bench with unused space that would be perfect to load up with a mini-plant gang of peperomia? Or turn a blank wall into a jaw-dropping green gallery for a ton of your leafy friends.

Don't forget to consider what you do in your kitchen. Do you have young children often taking up bench space with their creative endeavours? Perhaps a low, wide-reaching plant that needs a lot of space isn't a great idea. A tall vase filled with foliage from the garden could be a great alternative. Or maybe you have a rowdy dog that loves to play ball games bulldozing through the kitchen? Rather than using a tall plant stand, which may easily topple if knocked, think about choosing a shorter stand with a pot that nestles for added strength, and team it with a robust specimen like snake plant or a kentia palm.

Introduce varied heights in your plant styling by using trailing specimens like heartleaf philodendron on top of cabinetry. Layer the arrangement by pairing earthy ceramics with a chain of hearts or sculptural chain cactus on shelves.

Layer kitchen shelving with all manner of earthy-inspired ceramics, sculptures and a variety of plants like spider plant, lipstick plant, *Hoya heuschkeliana* and Chinese money plant.

THE BIG STUFF

Start by reviewing your kitchen on a macro scale. We're talking cabinetry, furniture, benches – all the big stuff. Ask yourself a few questions. What are these pieces made of? Are they natural materials? What shape are they? Are they textured? Do they have patterns similar to those you might find in nature? How about their colour? The goal here is to identify which pieces have a connection to nature in some way. And if they don't, think about how they can be tweaked or updated to bring more of the outside in.

Always find space for at least one or two plants in your kitchen. Plants help soften the harsh lines that are often prevalent in such a functional part of the home. Grape ivy, green exotica hoya and Chinese money plant are specimens that work well.

When working with an existing kitchen, you may be limited by budget or space, so it pays to get really creative. The big things are always harder to navigate but the pay-off can be huge. For instance, if your benchtop is synthetic and the budget allows, how about replacing it with natural stone or handmade terracotta tiles? Or, for a more budget-sensitive option, perhaps you could add a recycled timber butcher's block. If seating is an option, you could introduce vintage timber stools for a dose of the natural world.

Cabinetry plays a big role in a kitchen. If you're keen to introduce a natural vibe where there's currently none, look to re-paint or re-cover it with wallpaper that speaks to nature. Perhaps you can add rattan or natural fibre fronts to the doors or replace them with a natural alternative. Splashbacks are also worth reassessing. Maybe you could paint your splashback to reflect natural colours, re-tile it in a handmade ceramic or switch to stone, if you are so inclined.

Paint is your best friend when it comes to rejuvenating spaces and is especially useful when considering those large-scale elements in a kitchen. If you have stark white cabinetry that feels too cold, paint it! Boring beige walls? Paint them! Try updating in a shade of green, terracotta or charcoal grey. In fact, go for any earthy shade that echoes the outdoors and draws upon the natural world in a modern and tasteful way. If you're renting and aren't able to paint, try upcycling your kitchen furniture with a new look instead.

THE SMALL DETAILS

Next, work your way down to the smaller things like tableware, decor pieces and even crockery and utensils. Consider buying handmade ceramics for the kitchen, taking your time to build your collection. Each piece will have its own charm, and supporting small businesses and local artisans with your hard-earned dollars is very rewarding. Don't forget to get thrifty too. We often scour vintage markets and op-shops for tableware, ceramics, placemats, timber and brass bowls and a whole host of other eco-inspired pieces. It's a great way to contribute to a sustainable circular economy.

Plants for the kitchen

We spend so much time in our kitchens, it makes sense to have plants that perform well while also looking good. These are our top three multitaskers.

English ivy

This classic climbing and trailing plant may help filter out harsh toxins from common kitchen cleaning products and reduce the incidence of mould. With its winding tendrils, it has a softening effect that looks good trailing over shelves or on top of kitchen cabinetry, as well as suiting dining tables and even benchtops. Just be sure to keep the tendrils on the short side so they don't get in the way.

Spider plant

This popular plant is recognised as potentially reducing the effects of airborne household toxins, including formaldehyde, a chemical commonly found in products such as adhesives and grout, which are often present where there are tiled splashbacks and benches. With their pale green and white fronds, spider plants are a cute addition to any plant gang and they're super easy to propagate.

Aloe vera

Aloe vera is best suited to the sunniest spot in your kitchen. Not only is this attractive rosette-shaped plant one of the best performing air purifiers, it's also handy to have these beauties within arm's reach to use as a salve on any kitchen burns or nicks.

LIVING ROOM

The living room is often the largest room in the house.
It is also recognised as a place of rest and relaxation,
a sanctuary away from the bustling world outside.
In this room, where a big part of our lives is spent, we
can often find ourselves lingering for extended periods
of time. This might mean sitting by a window enjoying
a quiet moment, lounging on the sofa marvelling at
flickering flames from a fireplace or playing games
on the coffee table with friends or family.

Create a flowing sense of space by displaying indoor plants throughout your home. Not only will you increase your direct connection to nature as you see, touch and care for them, you will also compound your connection to the outdoors too.

Large plants like the kentia palm and umbrella tree can enhance any living space. Their height and organic nature soften minimal interiors and add depth and interest to the room.

After the practical tasks of the day are finished with, we often retreat to the living room in our comfiest threads – to curl up, read a book, watch television, spend time with loved ones or just potter about – whatever it is we find soothing and nurturing as the day comes to a close.

Given the array of furniture, textiles, homewares and accessories found in this part of the home, you can go to town layering nature-inspired finishes and materials.

HOUSEPLANTS

Layer the plant life and don't be afraid to make the most of any space you can. If you don't have garden views, this is even more important. There's usually space for at least one small plant, or a quiet corner or two that are perfect for luscious big plants. Look for other opportunities as well – we're talking coffee or side tables, a fireplace hearth or mantel, or even a media or drinks cabinet. If there's a horizontal surface, get a plant on it!

Plants can also be used to create green dividing walls, delineating zones within open-plan living spaces. They also create focal points and green views, both complementary biophilic design elements. Don't forget about taking your plant game vertical by utilising shelving and wall or hanging planters. Vertical surfaces are all fair game when it comes to styling with plants, and they will help to get the most out of your space.

While functionality may reign supreme in the kitchen or bathroom, in a living room you can afford to have a little extra fun. Within living spaces there's a lot more scope to play with biophilic decor items, objects and mementos, teaming them with plants to create green vignettes or tablescapes. Try pairing your favourite plant or two with coffee table books, sculptural objects, bowls and even crystals to create cute scenes grounded in nature.

THE BIG STUFF

Living rooms offer ample opportunity to build up natural materials and a lot can easily be incorporated into any home. Start by reviewing the big things in your space.

Plants should always be part of a living room. Here *Lepismium bolivianum*, umbrella tree and peppercorn plant offer a little sanctuary from the outside world.

Wall treatments

Wall treatments shouldn't be overlooked. If you have naturally crafted walls in your home already – think brick, timber or mudbrick – make the most of them in their original state so their organic and nuanced texture can be appreciated. Whether you're looking at a feature wall or the entire room, blurring the lines between indoors and out is the goal here, and there are a number of options to explore.

Unlike the kitchen or bathroom, in the living room you don't have to worry about condensation or having to choose hard-wearing surfaces. Wall treatments that exude a natural ambience and celebrate imperfections are ideal. Limewash (made from crushed limestone) or Venetian plaster (lime plaster with pigment) can add depth and character to a room with their chalky, one-of-a-kind texture that gets better with age. Wallpapers like grass cloth, jute or even cork can provide a literal dose of nature on your walls. Even wallpapers with prints that echo scenes from the outdoors such as trees, flowers or animals are great choices. Timber panelling is another possibility and is available in a myriad of styles and finishes.

Fern leaf cactus perched in a wall planter makes for show-stopping art.

Art comes in many forms. Take inspiration from nature and create your own modernist pieces, forage for roadside grasses or perhaps use indoor plants as living artworks, with all their sculptural glory.

If you want to keep to a smaller budget and scale, turn to paint. It's the easiest and most cost-effective way to make a big change in a room. Don't forget paint isn't just for walls. Perhaps you have built-in cabinetry that could do with an update in a fresh, earthy shade. Old joinery can also be modernised by getting a little creative. Rattan, wallpaper and timber panels can all be added to cabinet fronts to give them a new lease of life. Cabinetry that has already been painted might benefit from being stripped back to its original hardwood. Don't forget about the detailing either. Adding leather or brass handles, for instance, can have a big impact for a relatively small outlay.

Art is another element which should never be overlooked. If you want a quick fix and can't bear the thought of painting, adding artwork can be a simple way to change your space. There's a wonderful array of affordable plant-based art and prints available these days. Why not try creating your own modern graphic art using colours reflective of Mother Nature, or scouring vintage stores or flea markets for books that have pages you could frame. What about some vintage landscape paintings? Be on the lookout for mirrors, weavings and other wall adornments that evoke a sense of nature either in form, colour, pattern or texture.

Layering

With the walls taken care of, it's time to layer up the room. Start with your biggest pieces of furniture: sofas, chairs, cabinets and coffee tables. Are they crafted from a variety of natural materials? If not, is there a way to update them? Or can you add some smaller pieces of furniture to add a touch of nature in the room? Occasional tables and stools are a great way to do this. They take up minimal space and can be used in a number of ways – a dinner table for small children, extra seating when guests visit or a side table for your cuppa – and you can really have fun with your choice of finishes. Oak, cork, terrazzo, leather and limestone are just a sample of what's available. Wool, linen and other outdoor-inspired textiles can be used for larger furnishings like sofas, rugs, chairs and curtains to help round out the scene.

Include all manner of natural materials in your home to create a layered connection to nature. Try functional cork for stools, textured wool for carpets and upholstery, and rattan or willow for lampshades.

THE SMALL DETAILS

For utmost comfort, you can't go past cushions and throw blankets made from natural fibres (inside and out). Think wool, cotton or linen. Alpaca is a luxurious choice if the budget allows. Cushions, in particular, are a budget-friendly option that pack a big punch in a living room – they can add a pop of colour or pattern that a space needs. They also allow you to create a whole new look quickly if you want to refresh the room without having to replace all your larger furniture items.

Don't underestimate the smaller finishing touches you can add to really anchor the connection to the outdoors. Burn essential oils, candles or your favourite incense. Find scents crafted from nature that speak to you and have a grounding, calming effect.

Plants for the living room

We love styling up a storm in living rooms, which offer a multitude of spaces just calling out for a touch of greenery. From empty corners in need of a large statement plant to teacup-sized plants for coffee tables and trailing plants for shelving, opportunities present themselves at every turn.

Fiddle leaf fig

It's hard to go past this fig when you're after a large tree-shaped statement plant. Named after their fiddle-shaped leaves, these glossy green giants achieve spectacular heights when given the right growing conditions. Just remember that these guys love a bright filtered spot, moderate watering and a high ceiling.

Jade peperomia

An Ivy Muse favourite, this popular variety of peperomia has plump, deep-green leaves. Known for their easygoing nature, these bushy yet compact plants can add the perfect note of greenery to smaller surfaces like coffee tables and side tables.

Heartleaf philodendron

A low-maintenance philodendron with leaves shaped like hearts, there's a lot to love when it comes to this luscious trailing plant. Perfectly at home perched on top of shelving or on the edge of the entertainment console, these deep-green beauties grow long and luscious. Our tip: put these plants in a self-watering pot to lengthen the time between waterings.

BATHROOM

If you're like us and have young children, the bathroom might be the one place at home where you can catch a moment to yourself. (Well, sometimes at least!) Here we can restore internal harmony with quiet contemplation and invigorate ourselves, gathering up our reserves for another day in the world. This space is home to some of our most precious wellbeing rituals and plays an important role in a biophilic home. The bathroom is often the smallest and most plainly decorated room in a house, but there are still ways to make it biophilia-friendly without resorting to a complete overhaul.

HOUSEPLANTS

When it comes to adding greenery to a home, the bathroom should not be overlooked because it's actually the perfect habitat for indoor plants. Bathrooms provide a warm, humid environment, which mimics the more tropical conditions where the vast majority of indoor plants originate. Finicky indoor plants like ferns are at home here, as are other moisture-loving plants. Common complaints such as fronds with brown, dried tips suddenly become a thing of the past. An added bonus is that plant care becomes a breeze. Simply place your plants in the sink or bath (or taller plants under the shower) for a thorough drenching.

Common concerns like a bathroom with little natural light can be overcome by selecting the right plant from the get-go. A striking Zanzibar gem will thrive in low-light environments and welcome the humidity. A tiny bathroom? Get creative with your use of space. Why not perch a potted plant on a shelf, hang a planter above the bath or simply place a moisture-loving air plant in your shower accessories cavity? Even the smallest touch of greenery will add softness and warmth.

With all their hard surfaces, bathrooms can be stark and sterile in contrast to other rooms, as they are usually devoid of soft furnishings other than a towel or two. Plants can make a real difference, softening the look and feel and ultimately making it a more appealing place to be. Layer the plant life using small plants like hoya on vanities or windowsills, and add bigger options like umbrella trees in plant stands or rubber plants on the floor. If you can fit in a kentia palm or other tall tree like a fiddle leaf fig, go for it – they really help create a luxurious sanctuary vibe.

THE BIG STUFF

First up, assess. Where can you add layers, warmth and depth? Modern bathrooms often have clean straight lines so try to mix up the shapes and textures and lean towards organic, curvy forms that reflect those found in nature and that offer contrast. It's worth focusing efforts on elements that can be retrofitted into your bathroom without too much trouble.

Take a moment to check out your vanity. If it's constructed from synthetic materials can you replace the top with a more sensual, natural surface like marble or timber? How about the cabinetry? Can it be painted or sanded back to its original state, or new doors or pulls added instead?

Paint is a quick and relatively easy option to update walls and the ceiling. You can even paint the tiles if you're feeling daring, but make sure you choose tile paint which is purpose-made. Wall treatments including washable wallpaper could be a possibility, depending on how the home and room is decorated already. For example, adding cedar panels to a 70s bathroom could look great. Not only will it fit with the style of the era but the fragrant smell will engage your senses and increase the biophilic effect in the room.

Wall adornments that can withstand bathroom condensation are a big yes too. Think about hand-made ceramic sculptures, round or oval mirrors or art made from woven eco-materials such as grasses.

Furniture options in the bathroom are sometimes limited, but even the smallest spaces can benefit from a stool. Try timber for its durability, wondrous feel and scent. We love solid timber or cork stools in particular. They're a great way to add a sculptural chunk of the outdoors inside. In large bathrooms that can accommodate extra items of furniture – a chair or freestanding cabinet, for instance – choose pieces that include biophilic elements like timber and stone. And remember, even the smallest touches can have a big impact. A cedar bath mat, for example, is functional and represents a strong connection to nature. It also smells great!

Air plants are easy-care evergreen plants that love humidity. To ensure your air plants thrive, place them in a bright spot out of direct sun and mist them regularly (two to three times a week). In drier environments, soak your air plants in lukewarm water for ten minutes each week, then shake off any excess water.

THE SMALL DETAILS

Choose sensual textures for the bathroom that really engage the senses. Start with quality towels. Natural fibres like Egyptian cotton or linen are the most tactile and comforting – and well worth the splurge in our opinion. Linen and cotton towelling also make great bathrobes.

Mini monstera has a distinctive organic form that makes a striking addition to any bathroom.

Scent plays a big role in the bathroom. Be aware of the presence of artificial fragrances and pungent chemicals in your personal care and bathroom cleaning products. Try to use natural and plant-based formulations where possible – the aroma is usually more subtle and reflective of what you might experience outdoors.

Ideally, bathroom surfaces should be styled for functionality, but always make room for at least one or two plants. They provide a pleasant biophilic view while you're taking a bath or shower.

Plants for the bathroom

Turn your bathroom into your own personal oasis with the simple addition of greenery. Although it's often not the first room to spring to mind when considering indoor plants, it's worth noting that bathrooms naturally produce the perfect environment for the majority of indoor plants because of their warm and humid conditions.

Prayer plant

With its unusual colours and markings, the prayer plant is the perfect plant for bathrooms that are a little on the shady side. Known for thriving in low-light conditions, these humidity-loving plants are multicoloured wonders that have the ability to flower when provided with the perfect growing conditions.

Elephant ear

Often compared to a sculptural work of art, the elephant ear is easily recognisable by its distinct veined, leathery foliage. A visual delight, these plants thrive in bright, indirect light and humid conditions that mimic their tropical origins. Just remember to ensure that your vessel has adequate drainage, as these popular indoor plants won't tolerate waterlogged soil.

Air plants

While air plants can be a little 'needy' in other rooms in your home, requiring frequent waterings and misting, the humidity in bathrooms can take these plants from high-maintenance to easy-care in a heartbeat. With their ability to absorb vaporous moisture through the stoma in their leaves, you're literally watering these unique creatures every time you take a shower. Passive plant care at its finest!

BEDROOM

One third of our lives is spent in bed. Sleep is such an integral part of good health, and with more of us devoting less time to this act than ever before, it should be prioritised. Cocoon yourself in a layered, natural theme that encourages rest and relaxation. Sleep under sheets of cotton and linen. Experience the warmth of textured wool carpets. Restore your energy by surrounding yourself with a multitude of sensuous and comforting biophilic materials.

HOUSEPLANTS

The bedroom is the ultimate room of rest – a place to retreat and rejuvenate. It makes sense to surrender to nature here and 'green it up' for maximum effect. We love the idea of creating a multifaceted approach that cultivates a quiet, gentle energy. Focus on selecting plants that embody a soft, enveloping nature, such as the umbrella tree with its welcoming canopy, the coin leaf peperomia with its distinctive foliage or zigzag cactus with its trailing tendrils. Avoid plants that embody a more intense energy like snake plant or towering succulents with sharp edges.

Instead of using items that have a 'heavy' energy, such as concrete pots or ceramics in dark, solid colours, consider using 'softer' vessels for your plants. Woven baskets, or ceramic pots with lighter, more organic glazes, are perfect choices for the bedroom.

Wicker or rattan planters offer great contrast to smooth foliage, and their earthy timber shades complement most interior schemes.

Play around with the positioning too. If space allows, try clustering various plants in one arrangement to create visual interest. Arrange the display in tiers, with the tallest plant sitting at the back, then the medium-sized one, working your way down to the smallest. Plant stands also help to achieve staggered heights. The contrasting colours and shapes will work in synergy with grouped plants. They will soften each other's corners and blend various shades of green to form a serene scene, reminiscent of an outdoor experience.

Regardless of whether you have room for only a few plants or can go full jungle, indoor plants help alleviate modern stressors and are an integral component of any biophilic bedroom.

THE BIG STUFF

Bedrooms are often simple spaces, designed for functionality and comfort. When you want to increase the biophilic qualities of the room, your bed is an easy place to begin. Start with a comfortable bed that suits you, then add natural bedding materials such as linen, cotton or bamboo. Using linen in a striking colour can also be a relatively affordable and easy way to add impact without having to resort to painting walls or changing wall treatments. Don't be afraid to add lots of cushions and textured throws made from wool, cotton or other natural fibres to add depth and warmth to the bed. They can be the perfect finishing touch when it comes to styling it too.

Assess the other furniture in the room. Think about your side tables, dresser or shelves. What are they made from? Can they be tweaked or updated to be more biophilic? Adding items like plants, crystals and candles will increase the connection to nature without the need for more intensive DIY projects.

Curtains are important in the bedroom. Although it's great to be able to see outside easily, when it comes to sleeping, dark is best. Our circadian rhythm (and therefore quality of sleep) is hugely affected by external light sources such as street lights. Introduce a natural fabric here if possible. Linen teamed with block-out blinds can be a winning combo.

Keep lighting warm and inviting rather than stark and overpowering. Lamps are often the best way to create soft mood lighting that will help you feel safe and cocooned. Use downlights only if you have to and don't be afraid to use the dimmer.

A coffee table vignette complete with books, crystals and handmade objects sets the backdrop for an eye-catching *Hoya heuschkeliana*.

THE SMALL DETAILS

The bedroom provides a comfortable retreat where we can curl up when feeling worn out and in need of rest and recuperation. It makes sense to layer the smaller decor items here, especially those pieces that have sentimental meaning rather than just functionality. Organic ceramics, artworks and decor pieces like candles, books and trinkets all combine to comfort and nurture us daily.

Dedicate a portion of your bedroom to create your own sacred space for mindfulness, meditation and manifestation. Whether you choose to call it an altar or a place of reflection, this sacred space can feed the part of you that longs for connection to nature.

Plants for the bedroom

Plants have the ability to make us slow down and relax. Being in their presence often evokes a sense of calm and serenity – the perfect reason to ensure you add at least one, if not more, to the bedroom. When choosing plants for the bedroom, consider one of the following. The touch of greenery will make your room feel like a sanctuary.

Peace lily

One of our go-to plants and for good reason. As well as boasting the ability to filter toxins from the air, the peace lily also has potential sound-absorbing qualities due to its large, thick leaves – a great addition to any room where silence is golden. In a larger size, this plant's full form works well in a plant stand or on the floor, while smaller specimens look great on a windowsill or cabinet, where they can really stretch their leaves to the sky.

Bromeliad

Beautiful foliage and blooms are not all these plants have to offer. With their long, elegant fronds, bromeliads often appear to be reaching out to you with a warm welcome. Great for the novice indoor gardener, these vibrant plants have an easygoing temperament. Just remember to amp up the humidity and closely monitor watering needs. The rosette shape of most bromeliads creates an inner well through which these plants absorb water. Ensure the inner well is always full, using rainwater or filtered water, and flush out stagnant water every month or so.

Golden cane palm

These tall beauties are reminiscent of bamboo with their smooth golden trunks and narrow fronds. They are the perfect way to add a tropical vibe to your bedroom. Best suited to the more experienced indoor-plant enthusiast, these palms have a penchant for slightly acidic soil, frequent watering and feeding, and warm temperatures. Your attention and care will be reciprocated with a lengthy indoor life span of ten or more years.

RETREAT SPACE

Our homes offer us respite from the bustling world, so it's important to have at least one space – or better still, a room – where you can let go of the day and ground yourself surrounded by nature. This might mean relaxing on your bed among dream-worthy linens, nestling into a comfy old chair in a quiet spot, heading to your garden studio or rolling out your yoga mat in your own pocket-sized nook. The goal is to define a special spot within your home that makes you feel comforted, where you can disengage from everyone around you (yes, even the kids or housemates for a moment), rebalance your energy and restore yourself.

Time spent alone these days is like gold; we all need it to recharge mentally and physically. Taking time out for self-care rituals ultimately makes us more content when we do spend time with loved ones and friends, so it's a win–win situation for everyone. Do the things you enjoy here too. It might be painting, meditating, journalling or just sitting quietly. Any space set aside, no matter what the size, can elevate the activity and create a meaningful ritual.

HOUSEPLANTS

In contemporary homes, rooms often have multiple uses. For example, the living room may also serve as a home office or yoga studio, depending on the time of day. When space is at a minimum, it makes sense to get the most out of it. These spots within a home need houseplants too, but choosing the right ones requires some consideration.

Rather than having big, wide-reaching plants like a fruit salad plant that may get in the way of your downward dog and become a nuisance, especially as it grows bigger, incorporate small to medium plants like mistletoe cactus or devil's ivy, which are a lot more flexible with their placement. If possible, utilise shelving, wall and hanging planters to keep your houseplants up and out of the way while all that activity takes place.

Your retreat space should nurture and ground you, so choose plants that you really connect with. If you're restricted by less-than-ideal lighting conditions and your plants are suffering, try rotating them every few days into a brighter position or, as a last resort, use cut foliage instead of plants, replacing them every week or so with foraged finds from your garden or local area.

If you don't have the luxury of having a dedicated retreat space to call your own, get creative with your plant placement wherever you like to rest and take five. Even small touches of plant life can be effective, whether placed at bedsides, on walls, mantels or simply on the floor.

Rhipsalis elliptica, devil's ivy, variegated creeping fig and dragon tail plant

Krimson princess hoya

Grape ivy and air plant

Fashion a biophilic shelfie with a mix of materials, shapes and textures that echo those found outdoors. A bamboo trinket box, brass oil burner, handmade ceramic vessels and plants combine to create a layered vignette.

Layering

Your retreat space presents a perfect opportunity to surround yourself with various biophilic elements that speak to you. Whether you have a penchant for sculptural ceramics, antique wooden crafts or hand-loomed Moroccan wool rugs, prioritise and make room for the things that have the most meaning to you. Even if you are only working with a corner of a room, you still have the opportunity to layer up and have fun. Remember this spot is just for you so fill it with things from nature that make your heart sing.

Plants for your retreat space

Your retreat space is the perfect room (or nook) to plantify. Specimens with soft, round leaves or feathery foliage capture the gentle, nourishing energy that goes hand in hand with rest and reflection. As organic living organisms that are always growing and moving – never static – the plants in your sanctuary can encourage you to unblock stagnant energy present in your body and grow and move forwards in new ways.

Happy plant

A great option if your retreat space is low on natural light, this shade-loving plant exhibits personality plus! With its ribbon-like foliage and contrasting pale green variegation, the happy plant is sure to make you smile. A sturdy trunk supports this plant as it grows to new heights. Just ensure there is adequate drainage and only water when the soil is dry.

Boston fern

The Boston fern with its feathery fronds is a striking favourite of ours. While it requires considered placement – think a cool place with high humidity and indirect light – once you get into the flow of taking care of this lime green beauty, it will flourish and grow under your watchful eye.

Jade plant

A feng shui favourite believed to bring good luck and wealth to its owner, the jade plant is also admirable for its happy-go-lucky attitude. As succulents, these strong-rooted, evergreen plants require little water and attention. Their very presence is considered auspicious, bringing a positive, grounded energy to any room they are placed in.

05
Further reading

BIOPHILIA AND BIOPHILIC DESIGN

14 Patterns of Biophilic Design: Improving Health and Well-Being in the Built Environment, WD Browning, CO Ryan & JO Clancy, Terrapin Bright Green, LLC, 2014 <terrapinbrightgreen.com/reports/14-patterns/>.

The Biophilia Effect: A Scientific and Spiritual Exploration of the Healing Bond between Humans and Nature, Clemens Arvay, Sounds True, 2018.

Biophilia: The Human Bond with Other Species, EO Wilson, Harvard University Press, 1984.

Biophilia: You + Nature + Home: A Handbook for Bringing the Natural World into Your Life, Sally Coulthard, Kyle Books, 2020.

Biophilic Design: The Theory, Science, and Practice of Bringing Buildings to Life, Stephen R Kellert, Judith Heerwagen & Martin Mador, John Wiley & Sons Inc., 2008.

The Nature Principle: Human Restoration and the End of Nature-Deficit Disorder, Richard Louv, Algonquin Books, 2011.

The Practice of Biophilic Design, Stephen R Kellert & Elizabeth F Calabrese, 2015 <biophilic-design.com>.

Sensual Home: Liberate Your Senses and Change Your Life, Ilse Crawford, Quadrille Publishing, 1997.

INDOOR PLANTS

The Healing Power of Plants: The Hero House Plants That Will Love You Back, Fran Bailey, Random House UK, 2019.

Indoor Green: Living with Plants, Bree Claffey, Thames & Hudson, 2015.

Plant Style: How to Greenify Your Space, Alana Langan & Jacqui Vidal, Thames & Hudson, 2017.

Plantfulness: How to Change Your Life with Plants, Jonathan Kaplan & Julie Rose Bower, Laurence King Publishing, 2020.

NOTES

p. 20 'A landmark study in 1984 found that contact with nature': RS Ulrich, 'View through a Window May Influence Recovery from Surgery', *Science* 224(4647), 1984, pp. 420–1.

p. 61 'Many people feel drawn to curved lines': Paul J Silvia & Christopher M Barona, 'Do People Prefer Curved Objects? Angularity, Expertise, and Aesthetic Preference', *Empirical Studies of the Arts* 27(1), 2009, pp. 25–42; Moshe Bar & Maital Neta, 'Humans Prefer Curved Visual Objects', *Psychological Science* 17(8), 2006, pp. 645–8.

06
Plant care

COMMON NAME	SCIENTIFIC OR LATIN NAME	WATERING GROWTH PERIOD (WARMER MTHS)	WATERING REST PERIOD (COOLER MTHS)	LIGHT	HUMIDITY
African milk tree 'Royal Red'	*Euphorbia trigona* 'Royal Red'	💧	💧	⛅	〰
Air plants	*Tillandsia* spp.	See page 122	See page 122	⛅	〰〰
Aloe vera	*Aloe barbadensis* 'Miller'	💧	💧	⛅ – ☀️	〰〰
Aluminium plant	*Pilea cadierei*	💧💧💧	💧💧	⛅	〰〰
Begonia	*Begonia* spp.	💧💧	💧	⛅	〰〰
Bird of paradise	*Strelitzia reginae*	💧💧	💧	☀️	〰〰
Bird's nest fern	*Asplenium nidus*	💧💧💧	💧💧	⛅	〰〰
Blue star fern	*Phlebodium aureum*	💧💧💧	💧💧	☁️ – ⛅	〰〰
Boston fern	*Nephrolepis exaltata*	💧💧	💧💧	⛅ – ☀️	〰〰 – 〰〰〰
Bromeliads	Bromeliaceae	See page 139	See page 139	⛅	〰〰 – 〰〰〰
Bunny ears cactus	*Opuntia microdasys*	💧💧	💧	☀️	〰
Cast-iron plant	*Aspidistra elatior*	💧💧	💧	☁️ – ⛅	〰
Century plant	*Agave americana*	💧	💧	☀️	〰
Chain cactus	*Rhipsalis paradoxa*	💧💧	💧	⛅	〰〰 – 〰〰〰
Chain of hearts	*Ceropegia woodii*	💧	💧	⛅ – ☀️	〰 – 〰〰
Chinese money plant	*Pilea peperomioides*	💧💧	💧	☁️ – ⛅	〰〰
Christmas cactus	*Schlumbergera truncata*	💧💧💧	💧💧	☁️ – ⛅	〰〰
Coin leaf peperomia	*Peperomia polybotrya*	💧💧	💧	⛅	〰〰 – 〰〰〰
Devil's ivy	*Epipremnum aureum*	💧💧	💧	☁️ – ⛅	〰〰 – 〰〰〰
Donkey's tail	*Sedum morganianum*	💧💧	💧	⛅	〰 – 〰〰
Dragon tail plant	*Epipremnum pinnatum*	💧💧💧	💧💧	☁️ – ⛅	〰〰 – 〰〰〰
Dwarf umbrella tree	*Schefflera arboricola*	💧💧	💧	⛅	〰〰 – 〰〰〰
Elephant ear	*Colocasia* sp.	💧💧💧	💧💧	☁️ – ⛅	〰〰 – 〰〰〰
Emerald ripple	*Peperomia caperata*	💧💧	💧	⛅	〰〰

COMMON NAME	SCIENTIFIC OR LATIN NAME	WATERING GROWTH PERIOD (WARMER MTHS)	WATERING REST PERIOD (COOLER MTHS)	LIGHT	HUMIDITY
English ivy	Hedera helix	💧💧	💧💧	☁️ – ⛅	~
Fern leaf cactus	Selenicereus chrysocardium	💧	💧	☁️ – ⛅	~ – ~~
Fiddle leaf fig	Ficus lyrata	💧💧	💧	⛅ – ☀️	~ – ~~
Firesticks	Euphorbia tirucalli	💧	💧	☀️	~
Fruit salad plant	Monstera deliciosa	💧💧	💧	⛅	~
Giant bird of paradise	Strelitzia nicolai	💧💧	💧	☀️	~
Golden cane palm	Dypsis lutescens	💧💧	💧	⛅	~ – ~~
Grape ivy	Cissus rhombifolia	💧💧	💧	☁️ – ⛅	~
Green exotica hoya	Hoya carnosa 'Exotica'	💧💧	💧	⛅ – ☀️	~
Happy plant	Dracaena fragrans 'Massangeana'	💧💧	💧	⛅	~
Heartleaf philodendron	Philodendron scandens	💧💧	💧	☁️ – ⛅	~ – ~~
Hoya heuschkeliana	Hoya heuschkeliana	💧💧	💧	⛅ – ☀️	~
Hoya odetteae	Hoya odetteae	💧💧	💧	⛅ – ☀️	~
Hurricane cactus	Lepismium cruciforme	💧💧	💧💧	☀️	~ – ~~
Jade	Crassula ovata	💧💧	💧	☀️	~
Jade peperomia	Peperomia rotundifolia	💧💧	💧	☁️ – ⛅	~ – ~~
Japanese andromeda	Pieris japonica	💧💧	💧	⛅	~
Japanese aralia	Fatsia japonica	💧💧💧	💧	⛅	~
Kentia palm	Howea forsteriana	💧💧	💧	⛅	~ – ~~
Krimson princess hoya	Hoya carnosa 'Rubra'	💧💧	💧	⛅ – ☀️	~
Krimson queen hoya	Hoya carnosa 'Tricolor'	💧💧	💧	⛅ – ☀️	~
Lady palm	Rhapis excelsa	💧💧	💧	⛅	~ – ~~
Lepismium bolivianum	Lepismium bolivianum	💧	💧	⛅ – ☀️	~
Lipstick plant	Aeschynanthus radicans	💧💧	💧	⛅	~

COMMON NAME	SCIENTIFIC OR LATIN NAME	WATERING GROWTH PERIOD (WARMER MTHS)	WATERING REST PERIOD (COOLER MTHS)	LIGHT	HUMIDITY
Long leaf fig	*Ficus binnendijkii*	💧💧	💧💧	part sun	medium
Maidenhair fern	*Adiantum aethiopicum*	💧💧	💧💧	part sun	medium
Mexican snowball	*Echeveria elegans*	💧	💧	full sun	low
Mini monstera	*Rhaphidophora tetrasperma*	💧💧💧	💧💧	part sun	medium
Mistletoe cactus	*Rhipsalis baccifera*	💧💧💧	💧	shade	medium
Mountain aloe	*Aloe marlothii*	💧	💧	part sun – full sun	low
Never never plant	*Ctenanthe burle-marxii* 'Amagris'	💧💧	💧💧	part sun	medium – high
Painted lady	*Echeveria derenbergii*	💧	💧	full sun	low
Parlour palm	*Chamaedorea elegans*	💧💧	💧	shade – part sun	medium
Peace lily	*Spathiphyllum wallisii*	💧💧	💧	shade	medium
Peacock plant	*Goeppertia makoyana*	💧💧💧	💧💧	shade	medium
Peppercorn plant	*Piper nigrum*	💧💧💧	💧💧💧	full sun	medium – high
Philodendron 'Hope'	*Philodendron selloum* x hybrid	💧💧	💧💧	part sun	medium – high
Prayer plant	*Maranta leuconeura*	💧💧💧	💧💧	part sun	medium – high
Rabbit's foot fern	*Davallia fejeensis*	💧💧💧	💧💧	part sun	medium
Radiator plants	*Peperomia* spp.	💧	💧	part sun	medium
Rhipsalis elliptica	*Rhipsalis elliptica*	💧💧	💧💧	part sun	medium – high
Royal Hawaiian hoya	*Hoya pubicalyx*	💧💧	💧	part sun – full sun	medium
Rubber plant	*Ficus elastica*	💧💧	💧	part sun – full sun	medium
Rubber plant 'Tineke'	*Ficus elastica* 'Tineke'	💧💧	💧💧	part sun	medium – high
Snake plant	*Sansevieria trifasciata*	💧💧	💧	shade – part sun	low – medium
Snowdrop cactus	*Rhipsalis houlletiana*	💧💧💧	💧	shade	medium
Spanish moss (air plant)	*Tillandsia usneoides*	See page 122	See page 122	part sun	medium
Spath sensation	*Spathiphyllum* 'Sensation'	💧💧	💧	shade	medium

COMMON NAME	SCIENTIFIC OR LATIN NAME	WATERING		LIGHT	HUMIDITY
		GROWTH PERIOD (WARMER MTHS)	REST PERIOD (COOLER MTHS)		
Spider plant	*Chlorophytum comosum*	2 drops	1 drop	Filtered – Bright	Low – Moderate
Spurge	*Euphorbia* spp.	2 drops	2 drops	Filtered – Bright	Low
String of pearls	*Senecio rowleyanus*	2 drops	1 drop	Filtered – Bright	Low
Swiss cheese vine	*Monstera adansonii*	2 drops	2 drops	Filtered	Moderate – High
Tassel fern	*Huperzia nummulariifolia*	3 drops	2 drops	Filtered	Moderate
Umbrella tree	*Schefflera actinophylla*	2 drops	1 drop	Filtered	Moderate – High
Variegated creeping fig	*Ficus pumila* 'Variegata'	3 drops	2 drops	Filtered	Moderate
Variegated inch plant	*Tradescantia fluminensis* 'Variegata'	3 drops	1 drop	Filtered – Bright	Moderate – High
Weeping fig	*Ficus benjamina*	2 drops	1 drop	Filtered	Moderate
White ghost cactus	*Euphorbia lactea* 'White Ghost'	1 drop	1 drop	Filtered – Bright	Moderate
Zanzibar gem	*Zamioculcas zamiifolia*	1 drop	1 drop	Medium – Filtered	Moderate
Zebra plant	*Aphelandra squarrosa*	2 drops	1 drop	Medium – Filtered	Moderate – High
Zigzag cactus	*Epiphyllum anguliger*	1 drop	1 drop	Medium – Filtered	Moderate – High

INDEX LEGEND

Symbol	Name	Description
💧	One Drop	Water sparingly, allowing the mix to nearly dry out between waterings.
💧💧	Two Drop	Water moderately, allowing the top 3 cm to dry out between waterings.
💧💧💧	Three Drop	Water generously when surface of mix is dry.
☀	Bright Light	This plant thrives in bright, indirect sunlight and can tolerate periods of direct sunlight.
⛅	Filtered Light	This plant prefers bright, filtered sunlight but should not be placed in direct sunlight.
☁	Medium Light	This plant can tolerate shady conditions.
∼	Low Humidity	This plant prefers a dry atmosphere between 10 and 40% humidity. Dehumidifiers can be used.
≈	Moderate Humidity	This plant prefers a moderate amount of moisture, between 40 and 60% humidity. Most homes sit at this level, but check the humidity with a hygrometer if you are concerned.
≋	High Humidity	This plant prefers a high moisture content, with humidity higher than 60%. Mist the leaves of this plant every morning or sit the pot on a pebble tray.

ACKNOWLEDGEMENTS

A heartfelt thanks to everyone who helped make this book possible. It's been a project very close to our hearts and one that's so satisfying to see come to life. To our supporters near and far, this one is for you!

To Paulina, Elise, Ngaio and the team at Thames & Hudson, thank you for your belief in this book.

Huge thanks to the homeowners who opened their doors and welcomed us so warmly: Kate and Haslett from Coco Flip; Janine from White Salt Photography (Ferny Creek House, interior design by Haus of Hanem); Lauren and family from Lauren Egan Design; Brad and Sarah from Nicholls Design; the owners of Yarra Valley House (interior design by Chelsea Hing); designer Sarah Shinners; Ross Farm (interior design by Studio Moore); and Thomas Denning. It was such an honour to shoot your homes, and we're grateful for your kind hospitality.

To our friends and collaborators – the team at Whiting Architects (Forest Lodge project), Loretta (Hey Lenny), David Windsor, Amelia Stanwix, the team at Framing To A T, and all the designers, artists and fellow creatives whose work helped frame this book – thank you.

Big thanks also to our long-time collaborator Annette O'Brien, for helping to translate our vision for *Nature Style* onto paper in such a beautiful way.

Last but by no means least, thank you to our families and friends, who rallied around us during this project – we couldn't have done this without you. Special thanks to our partners for the endless cups of tea, and their support and patience. We love you.

And to our children, we dedicate this book to you.

First published in Australia in 2021
by Thames & Hudson Australia Pty Ltd
11 Central Boulevard, Portside Business Park
Port Melbourne, Victoria 3207
ABN: 72 004 751 964

First published in the United Kingdom in 2022
by Thames & Hudson Ltd
181a High Holborn
London WC1V 7QX

First published in the United States of America in 2022
by Thames & Hudson Inc.
500 Fifth Avenue
New York, New York 10110

Nature Style © Thames & Hudson Australia 2021

Text © Alana Langan and Jacqui Vidal 2021
Images © Annette O'Brien 2021

p. 9: image of authors by Amelia Stanwix

24 23 22 21   5 4 3 2 1

The moral right of the authors has been asserted.

All rights reserved. No part of this publication may be reproduced or transmitted in any form or by any means, electronic or mechanical, including photocopy, recording or any other information storage or retrieval system, without prior permission in writing from the publisher.

Any copy of this book issued by the publisher is sold subject to the condition that it shall not by way of trade or otherwise be lent, resold, hired out or otherwise circulated without the publisher's prior consent in any form or binding or cover other than that in which it is published and without a similar condition including these words being imposed on a subsequent purchaser.

Thames & Hudson Australia wishes to acknowledge that Aboriginal and Torres Strait Islander people are the first storytellers of this nation and the traditional custodians of the land on which we live and work. We acknowledge their continuing culture and pay respect to Elders past, present and future.

ISBN 978-1-760-76110-3
ISBN 978-1-760-76235-3 (U.S. edition)

A catalogue record for this book is available from the National Library of Australia

British Library Cataloguing-in-Publication Data
A catalogue record for this book is available from the British Library

Library of Congress Control Number 2021936880

Every effort has been made to trace accurate ownership of copyrighted text and visual materials used in this book. Errors or omissions will be corrected in subsequent editions, provided notification is sent to the publisher.

Design: Ngaio Parr Studio
Editing: Brigid James
Printed and bound in China by 1010 Printing International Limited

FSC® is dedicated to the promotion of responsible forest management worldwide. This book is made of material from FSC®-certified forests and other controlled sources.

Be the first to know about our new releases, exclusive content and author events by visiting
thamesandhudson.com.au
thamesandhudson.com
thamesandhudsonusa.com

Artwork credits

p. 2: photograph by Cathy Marshall; **pp. 6–7, 18, 100**: weaving, Hanne Ibach; **pp. 6–7, 71**: artwork by Sophie Moorhouse Morris; **pp. 8, 114–15**: artwork by Kayleigh Heydon; **p. 27**: timber wall sculpture by Mike Nicholls; **pp. 36, 64**: artwork by Guy Maestri; **pp. 39, 102**: artworks on wall by Nora Aoyagi and Tim Jones; **pp. 48, 53, 106, 108**: artwork by Sarah Shinners; **pp. 80, 92–3**: artwork by Stacey Rees; **p. 84**: artwork by Hannah Nowlan; **p. 102**: artwork on mantelpiece by Gabrielle Jones; **pp. 139, 141**: artwork by Lyn Turner.